Abdelaziz HADDAD

Papillomavirus and the p53 protein

Abdelaziz HADDAD

Papillomavirus and the p53 protein

Human papillomaviruses (HPV) are among the most widespread sexually transmitted infections.

Imprint

Any brand names and product names mentioned in this book are subject to trademark, brand or patent protection and are trademarks or registered trademarks of their respective holders. The use of brand names, product names, common names, trade names, product descriptions etc. even without a particular marking in this work is in no way to be construed to mean that such names may be regarded as unrestricted in respect of trademark and brand protection legislation and could thus be used by anyone.

Cover image: www.ingimage.com

This book is a translation from the original published under ISBN 978-620-6-71979-3.

Publisher:
Sciencia Scripts
is a trademark of
Dodo Books Indian Ocean Ltd. and OmniScriptum S.R.L publishing group

120 High Road, East Finchley, London, N2 9ED, United Kingdom
Str. Armeneasca 28/1, office 1, Chisinau MD-2012, Republic of Moldova, Europe
Printed at: see last page
ISBN: 978-620-7-98752-8

SUMMARY

Human papillomaviruses (HPV) are among the most widespread sexually transmitted infections: around 70-80% of the sexually active population will come into contact with these viruses during their sexual lives. There are different types of HPV virus. Certain types of virus cause condyloma (HPV-6 and 11 strains), also known as genital warts or cockscomb. These warts, single or in groups, generally develop in the ano-genital area (anus, perineum, penis, vulva) and are extremely contagious. They do not develop into cancer, but can be physically and psychologically embarrassing. Other types of HPV (HPV-16 and 18) cause precancerous lesions and can lead to cancers of the cervix, vulva, anus, penis and throat. However, there are vaccines that protect everyone against certain types of HPV. The p53 protein is a transcription factor that plays an important role in cancer. Discovered in 1979, it binds to DNA and promotes the expression of genes that repair cellular damage. The p53 protein controls the cell cycle and interacts with dozens of genes, Viral oncogenesis is a multi-stage process that depends largely on the complex interaction between viruses and host factors. Oncoviruses are capable of subverting cellular signalling machinery and metabolic pathways and exploiting them for infection, replication and persistence. Several viral oncoproteins are capable of functionally inactivating the tumour suppressor p53, causing deregulated expression of numerous genes orchestrated by p53, such as those involved in apoptosis, DNA stability and cell proliferation. So, when the cell is damaged, depending on the severity of the situation, activation of the p53 protein may aim to : either to control cell growth: p53 can block cell growth and promote cell repair, or to induce cell suicide (programmed cell death or apoptosis), to prevent the cell from becoming a tumour cell.

DEDICATION

ACKNOWLEDGEMENTS

First of all, I would like to thank GOD, the almighty and merciful, who has helped me and who gave me the patience and courage during these two years of study and the strength to accomplish this formidable task.

I would like to express my respect for my thesis supervisor, Professor SALIHA CHBICHEB, for her support, invaluable advice and encouragement. She was able to guide me in the construction and writing of my academic work. Also for the kindness and patience she showed me during my work and for the help and time she devoted to this end-of-study project.

despite his many responsibilities.

I would like to thank the members of the jury for the honour you have bestowed on me by agreeing to judge this work. Please accept the expression of my deep and respectful gratitude.

I would also like to thank my parents for the unconditional support they have shown me since I set out on my career path. Thank you for the financial and moral support, psychological and material. If I'm here today, it's thanks to you.

I would like to thank my friends and fellow students for their presence throughout these years. two years of study. Their support during the writing of this dissertation, their criticism and their proofreading were invaluable.

TABLE OF CONTENTS

INTRODUCTION

The term papillomavirus derives from the two Latin words "papilla", meaning "pimple, nipple", and "oma", meaning "tumour" as a whole. Until the end of the 19th century, warts were described and their contagious nature suspected, but their origin had not yet been established. In 1907, an Italian doctor by the name of Giuseppe Ciuffo demonstrated for the first time the viral origin of human warts by hetero-inoculation of an ultrafiltrate from a wart. HPV was first described in electron microcopy by Strauss et al in 1949.

Human papillomaviruses are viruses belonging to the Papillomaviridae family(1). To date, of just over 200 human papillomavirus genotypes identified, 120 have been fully sequenced, designated by number according to their historical chronological order of discovery (HPV 1 to HPV 120) (2).

According to the World Health Organisation (WHO), Human Papilloma Viruses (HPV) are responsible for a wide range of diseases, including precancerous lesions of the cervix which can progress to cancer, causing major public health problems worldwide (1).

The human papillomavirus (HPV) is the most common viral infection of the reproductive system. Most sexually active men and women will be infected at some point in their lives, and some may be infected more than once. The critical period of infection for both men and women is at the very beginning of sexual activity. HPV is transmitted during sexual intercourse, even if there is no penetration. Skin-to-skin genital contact is a well-known mode of transmission (3). HPV infection is the most common sexually transmitted infection in the world (4, 5). 75% of women will have been in contact with this virus during their sexual life. (6). Prevalence varies between regions of the world, with a maximum rate of 26% in Africa and a minimum rate of 8% in Asia (7).Certain types of human papillomavirus (HPV) are probable etiological agents of cancer and its precursors. Viral infection and the expression of certain viral genes appear to be necessary but not sufficient factors in tumour transformation. Advances in molecular biology are shedding light on the mechanisms by which these viruses contribute to the development of cancer. Only the E6-E7 oncoproteins of high-risk papillomaviruses (HPV 16-18) bind specifically and with high affinity to the p53 and pRb proteins. This binding disrupts the cell cycle and leads to chromosomal instability. This event is probably the starting point for the integration of viral DNA into the host cell genome. These biological changes, in relation to morphological events and colposcopic aspects of dysplasia, appear to be very important stages in tumour progression (8).

The p53 protein was discovered in 1979 by Linzer and Levine (9). In 1989, Baker (10) highlighted its anti-oncogenic role. Between 1993 and 1996, more than 4,300 research studies were published on this protein. Its predominant role in cell cycle control and protection against proliferation in response to genotoxic stress has earned it the nickname of "genome guardian" or "tumour suppressor".The cellular protein p53 is a phosphoprotein found in vertebrates. It is almost certainly involved in cell cycle regulation. In vivo, p53 is able t o associate with various viral and cellular proteins. The p53 protein is a major transcriptional factor in genome protection. Commonly referred to as a "tumour suppressor" and "genome guardian", this protein is capable of activating and repressing genes involved in important biological processes. p53 is one of the most studied markers in oncology. P53 is one of the keys to cell regulation under genotoxic stress, and its response can take the form of apoptosis, senescence or cell cycle arrest, by activating DNA repair enzymes. For these reasons, its alteration leads to a possible step towards carcinogenesis. Diagnosis can be made by biopsy, looking for koilocytes, cells specific for HPV infection, represented by an infected keratinocyte, a large, round, vacuolated, clear cell with a central nucleus that is often pycnotic. There is often dyskeratosis, which is not specific for HPV. HPV testing can also be carried out using viral DNA research and genotyping (PCR, Southern-Blot, in situ hybridisation) (11). There is no cure for HPV, but certain strains of the virus can be vaccinated against. Regular use of condoms can reduce but not eliminate the risk of contracting or transmitting HPV during sexual relations.

I. GENERAL PROPERTIES

1. Classification :

Originally grouped together in the Papovaviridae family, papillomaviruses now form the Papillomaviridae family, which includes a large number of viruses that specifically infect various species of mammals and birds. To date, of just over 200 human papillomavirus genotypes identified, 120 have been fully sequenced, designated by number according to their historical chronological order of discovery (HPV 1 to HPV 120) (13).

Despite their identical structure, the degree of genomic homology between the different HPVs is only 40% (14).

The classification of HPVs is correlated with their tropism and pathogenicity. HPVs are classified according to their tissue tropism (cutaneous or mucosal) and their oncogenic potential (low risk or high risk) (15).

⤴ Tissue tropism classification

A distinction is made between HPV with cutaneous tropism and those with mucosal tropism. However, certain types of HPV can belong to both categories (Table I).

Table I: Classification of HPV according to tropism

Tropism	Types
Skin	1, 2, 4, 5, 8, 9, 12, 14, 15, 17, 19, 20, 21,
	22, 23, 25, 27, 36, 37, 38, 41, 47, 48, 49, 50,
	57, 60, 63,65, 75, 76, 80, 88, 92, 93, 95, 96
Mucous	6, 11, 13, 16, 18, 26, 30, 31, 32, 33, 34,
	35, 39, 42,44, 45, 51, 52, 53, 54, 56, 58, 59,
	61, 62, 66, 67, 68, 69, 70, 71, 72, 73, 74, 81,
	82, 83, 84, 85, 86, 87, 89, 90
Mixed	3, 7, 10, 28, 29, 40, 43, 78, 91, 94

⥥ Classification with oncogenic potential

There are around 40 types that infect the body's mucous membranes, such as the epithelium of the anogenital and oral tracts. Of these, 12 genotypes are officially recognised as HR-HPV (high risk oncogenic) on the basis of their frequency of association with malignant tumours: genotypes 16, 18, 31, 33, 35, 39, 45, 51, 52, 56, 58 and 59, according to the latest classification published by the World Health Organisation's International Agency for the Prevention of Cancer. Cancer Research Agency (IARC) (17). There are also 13 other genotypes classified as probably carcinogenic (HPV-68) or possibly carcinogenic (HPV-26, 30, 34, 53, 66),67, 69, 70, 73, 82, 85, 97). The genotypes (6,11,42,43,44,54,61,70,72,81) have a low oncogenic risk (LR-VPH) and are associated with benign lesions of the anogenital area called condyloma acuminata (genital warts), papillomas of the oral or respiratory cavity or conjunctiva, and low-grade squamous intraepithelial lesions (LSIL) of the cervix (18- 19).

Table II: Classification of anogenitally-tropic HPVs according to their potency oncogene

High oncogenic risk	16,18,31,33,35,39,45,51,52 ,56,58,59
Probable high risk	26,53,66,73,82
Low oncogenic risk	6,11,40,42,43,44,54,61,79,72

2. Structure and organisation of the genome

The papillomavirus viruses are a viruses belonging à the family Papillomaviridae(1).

These are small, naked viruses, 45-55 nm in diameter, with a cubically symmetrical capsid made up of 72 capsomeres in an icosahedral structure (Figure 1). Their genome consists of a circular double-stranded DNA molecule of around 8,000 base pairs (Figure 2), whose viral protein-coding sequences are grouped together on a single strand with different overlapping open reading frames (ORFs).HPVs have a common genetic organisation in three regions (Figure 3): a non-coding regulatory region LCR (Long Control Region) or URR (Upstream Regulatory Region), an E (Early) region coding for regulatory proteins or proteins involved in DNA replication (E1 to E7) and an L (Late) region coding for capsid proteins (L1 and L2) (21).

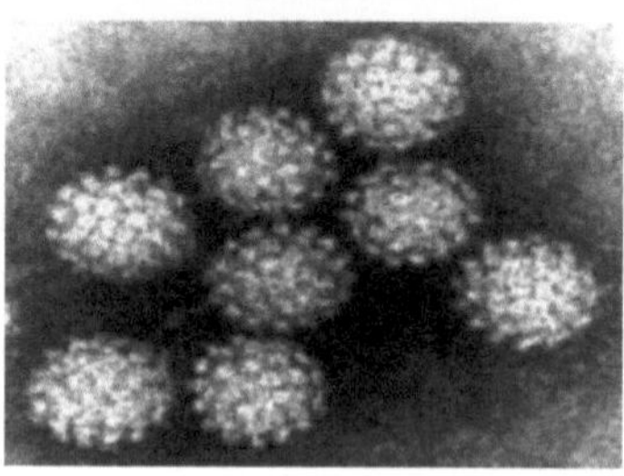

Figure 1: Structure of the viral capsid of papillomaviruses using an electron microscope

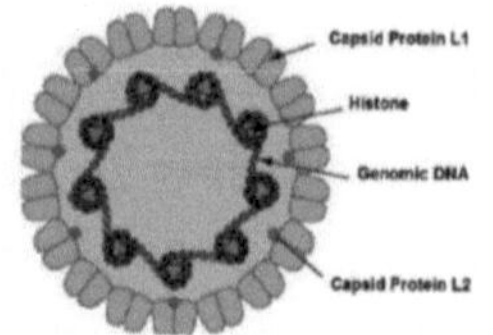

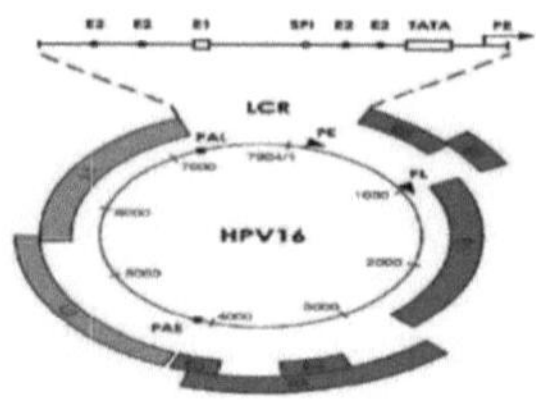

Figure 2: Structure of the papillomavirus
PE: Early promoter or p97 PL: Late promoter or p670 PAE: Early polyadenylation PAL: Late polyadenylation
Figure 3: Organisation of the HPV16 genome

All HPVs have a similar genomic organisation, which is divided into three regions (Figure 4).

An early E region, representing 45% of the genome, contains eight open reading frames (E1 to E7) encoding non-structural proteins known as functional proteins. These are involved in extra-chromosomal replication of the viral genome and its transcription (E1, E2), maintenance of the genome in episomal form (E2), maturation of viral particles and disorganisation of the cytoskeleton (E4), the processes of cell transformation (E5, E6 and E7) and cell immortalisation (E6 and E7).

A late L reading region: representing approximately 40% of the genome and composed of two ORFs (L1 and L2) coding for capsid structural proteins expressed only in well-differentiated cells during productive infections. These proteins enable final assembly of the virus and stabilisation of the capsid (24).

A non-coding region : LCR (long control region) or URR (upstream regulatory region): corresponding to the remaining 15% of the genome, this is highly variable and contains the promoters of early genes as well as transcripion regulatory sequences.

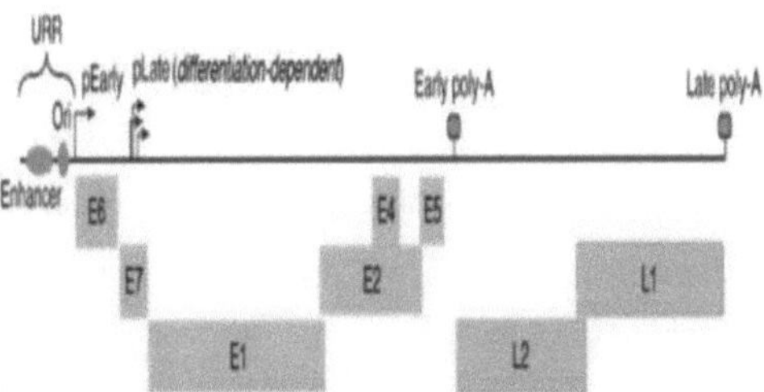

Figure 4: Genetic organisation of Alphapapillomaviruses

❖ **Early viral proteins E (Early):**

The E1 protein is required for viral DNA replication. Comprising 600 to 650 amino acids (68 to 75 kDa), it is the only enzyme produced by the virus. It has a helicase function, an activity that separates the two strands of DNA at the site of origin of replication, thereby promoting the establishment of a replication initiation complex, and an ATPase activity, both essential for viral DNA replication. The role of E1 in replication requires synergistic coupling with the E2 protein. In fact, E1 has no specific affinity for DNA and it is its association with the E2 protein that enables it, via the formation of E1-E2 heterodimers, to bind preferentially at the origin of replication to exert its helicase activity. A mutation in the E1 binding site (E1BS), or mutations in the E1 and/or E2 proteins are accompanied by a reduction or even a halt in viral replication. The E2 protein, made up of around 400 amino acids (50 kDa), is involved in both replication and modulation of viral transcription. As mentioned above, activation of replication requires a synergistic interaction with the E1 protein. The E2 protein also acts as a trans-inhibitory factor by binding to sites near the TATA box of the HPV16 p97 and HPV18 p105 promoters, causing steric hindrance at the transcription initiation site and thus repressing the expression of oncoproteins E6 and E7. It also plays a negative regulatory role in cell transformation and apoptosis. The E3 protein is only produced by very rare HPVs and its function remains unknown. The E4 protein, which is encoded by a highly diverse region of the genome, is a cytoplasmic protein that modifies the structure of keratin. The main E4 protein (17kDa) is translated from an E1^E4 mRNA. Although encoded by an early gene, E4 is expressed later and in abundance. Its expression occurs at approximately the same time as amplification of the viral genome, thus preceding that of the late L1 and L2 proteins. The E4 protein is then cleaved into multimeric structures that bind cytokeratins and promote destruction of the cytoskeleton, particularly in the superficial layers of the infected epithelium. In this way, the E4 protein is involved in the production of viral particles, facilitating encapsidation of the genome and promoting the diffusion and release of accumulated virions by destroying the network of cytokeratin filaments.

The E5 protein is a small hydrophobic protein of 8 to 10 kDa, located in cellular endomembrane systems. It is one of three oncoproteins encoded by the virus and is expressed in the precancerous stage but not generally in cancerous lesions. It is mainly expressed by oncogenic HPV or HR. Its functions are not perfectly understood. However, it is thought to be involved in viral DNA replication and transformation. The HPV16 E5 protein binds to the 16kDa sub-unit of the vacuolar ATPase H+, which interferes with the acidity of endosomes and consequently increases the recycling of EGF (Epidermal Growth Factor Receptor) and PDG (Platelet-Derived Growth Factor) receptors to the plasma membrane, as well as their activity. In addition, the E5 protein reduces recognition of the infected cell by the host immune system. It interacts with the MHC I heavy chain in the endoplasmic reticulum and Golgi apparatus, reducing its expression on the cell surface and therefore the presentation of viral antigens.

The E6 and E7 proteins play a key role, as they are required for the productive viral cycle and cell transformation induced by HR HPV. Their expression is partially controlled by the E2 protein. They are the only viral proteins constantly expressed in cervical cancer. They interact with numerous proteins and cell signalling and regulatory pathways. One of the main targets of the E6 protein in HPV HR is the p53 protein, which has a negative regulatory function in the cell cycle. In the event of DNA damage, it can induce cell cycle arrest, allowing repair of the damaged DNA. If the damage is too extensive, the p53 protein can induce apoptosis, preventing the spread of these alterations to subsequent generations of cells (Figure 5). The E6 protein of HR HPVs induces the degradation of p53 by the proteasome, whereas that of BR HPVs only reduces its activity.

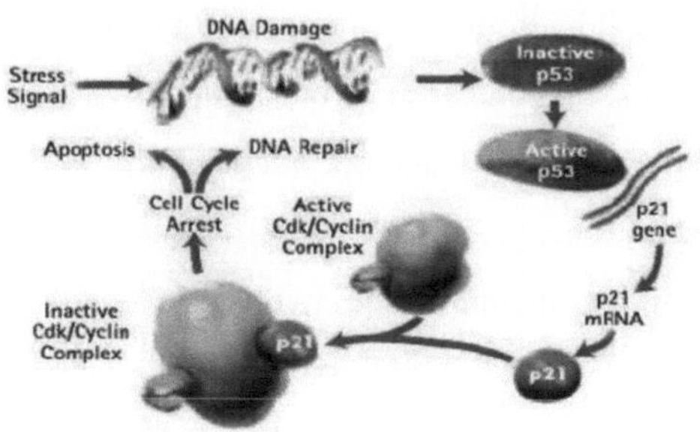

Figure 5: Physiological roles of the p53 protein

The function of the E7 protein in the productive viral cycle is to promote the transition from G1 to S phase, enabling genome replication in suprabasal cells. The E7 protein of HPV HR interacts with the pRb protein, whose main role is to block the replication of the genome in the suprabasal cells. passage of the cell from G1 to S by binding to the transcription factor pRb. E7 promotes its binding to calpain, which partially degrades pRb, causing it to be degraded by the proteasome. The joint action of these two oncogenic proteins is therefore part of the mechanism of carcinogenesis, since they are involved in the loss of control of cell cycle regulation, immortalisation, cell transformation and maintenance of the transformed phenotype.

Table III: Viral proteins and their main functions

Viral proteins	Features and functions
E1	Helicase. Controlling viral replication
E2	Negatively regulates the early gene promoter and, with E1, viral DNA replication
E4	May participate in the release of viral particles by destabilising the cytokeratin network
E5	Stimulates mitotic signals from growth factors
E6	Inactivates numerous cellular proteins, including p53. Major viral oncoprotein
E7	Inactivates many cellular proteins, including pRb. Major viral oncoprotein
L1	Major capsid protein. Found in directed vaccines against HPV
L2	Minor capsid protein. Promotes encapsidation of viral DNA

The L (Late) region codes for the late capsid proteins L1 and L2 (Table III).

❖ Late viral proteins L (Late) :

The L1 protein is the major capsid protein. Its structure is highly conserved between papillomaviruses, and it carries genus-specific antigens and certain type-specific antigens. Its ability to self-assemble to form viral like particles (VLPs) forms the basis of vaccines. VLPs share conformational epitopes with the native capsid and are highly immunogenic, yet non-infectious. They are also a source of antigens for the development of ELISA serological tests, which are not yet on the market. The L2 protein is the minor capsid protein: it is able to bind to viral DNA and position it correctly within the capsid. It therefore enables the virus to be assembled and the capsid to be stabilised in association with the L1 protein.

3. Viral cycle and Multiplication

HPVs are species-specific viruses with a particularly marked tropism for squamous epithelia, which are the only ones permissive to infection. These squamous epithelia are located in the skin (epidermal layer), the upper aerodigestive tract (oral cavity, pharynx, oesophagus, upper half of the larynx), the anal canal and the genital tract. HPV infects the epithelium by penetrating micro-lesions in the basal layer cells. In the cervix, this penetration is facilitated at the junction between the squamous epithelium of the ectocervix and the glandular epithelium of the endocervix, which is particularly vulnerable.

3.1. The viral cycle (28-29) :

a. Encounter with the target cell (Figures 6 and 7):

It has been observed in vivo that, before binding to keratinocytes, virions bind to the basement membrane via proteoglycans (mainly heparan sulphate). Heparan sulphate proteoglycan is a receptor found in the extracellular matrix and on the surface of many cells. Virions bind to heparan sulphate proteoglycan 1 (HSPG1) in the extracellular matrix via their L1 capsid protein. Laminin 5, present in the extracellular matrix, has also been shown to play a role in this binding, but to a lesser extent. The virions were then transferred to another receptor, heparan sulphate proteoglycan 2 (HSPG2), present on the surface of keratinocytes. This interaction would result in a conformational change leading to exposure of the N-terminal part of the L2 protein and the appearance of the furin cleavage site. This is followed by proteolytic cleavage, which again leads to a conformational change that reduces the affinity of the furin to the keratinocytes. for the heparan sulphate proteoglycan receptor or to give rise to a new binding site resulting in transfer to a second receptor leading to endocytosis. Some hypotheses suggest that alpha 6 integrin is the second receptor, but these hypotheses are controversial.

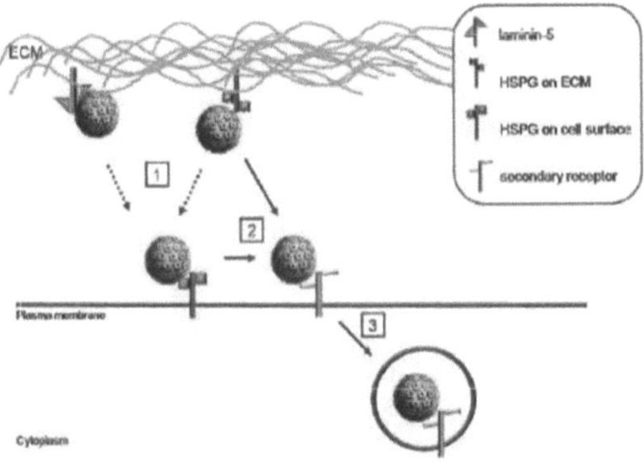

Figure 6: Mechanisms of cell entry by human papillomaviruses

b. Virus enters the cell :

HPV enters through slow endocytosis, the mechanism of which depends on the type of HPV. HPV 16 and 58 have an endocytosis mechanism via clathrins, whereas HPV 31 involves caveolae. The virus then decapsidises by breaking the bisulphide bridges between the capsomeres, migrates to the nucleus and releases the viral genome. The genome can remain in a latent state in episomal form or integrate the cell's genome.

c. Expression and amplification of the viral genome :

Once the viral genome has been released, viral proteins are expressed and the genome is replicated to form new virions.

d. Assembly of virions and their release by desquamation.

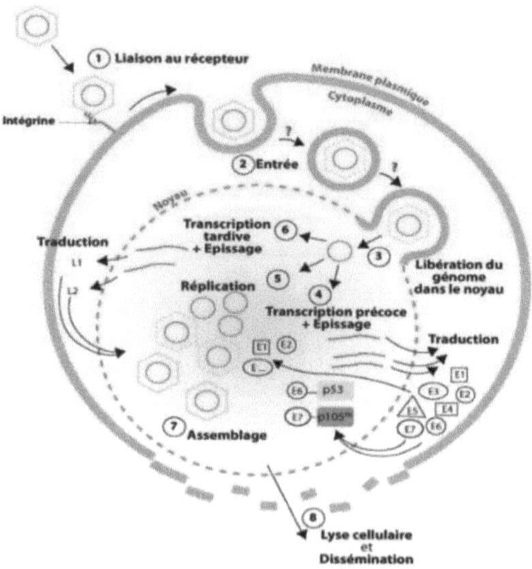

Figure 7: Human papillomavirus viral cycle

3.1. Viral multiplication (32, 33,34,35, 36,37,38)

The human papillomavirus has a narrow host specificity. It infects the germ cells of squamous epithelium of the skin and mucous membranes. This virus will go through three replicative phases (Figure 8).

a) Productive infection (Figure 9) :

If there is a breach in the epithelium, the virus will take advantage of this to enter the epithelium. basal cells of the epithelium and express its early genes. This leads to an expansion clonal proliferation of infected cells, leading to a benign cell proliferation called papilloma. Subsequently, the germ cells differentiate and migrate towards the keratinised surface layers. This is the late phase of the cycle when the L1 and L2 genes responsible for capsid synthesis are expressed. Viral particles are formed and released when the keratinocytes slough off. Viral multiplication is therefore closely correlated with cellular differentiation. This cellular differentiation is at the root of the cytopathic effect known as koilocytosis. Koilocytosis is the formation of abnormal cells, koilocytes, which have a very large, peripheral nucleus and irregular chromatin (Figure 8).

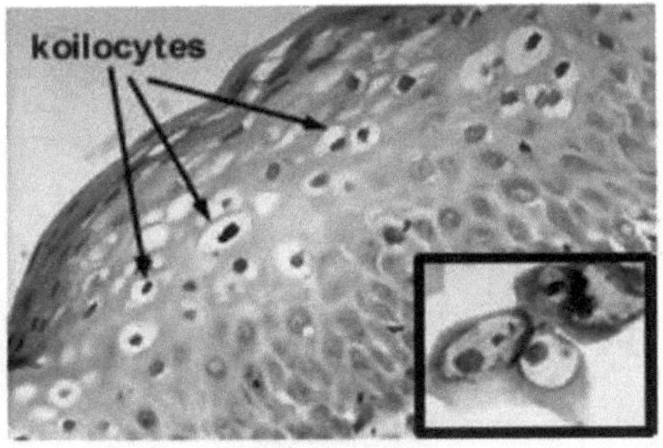

Figure 8: Koilocytes

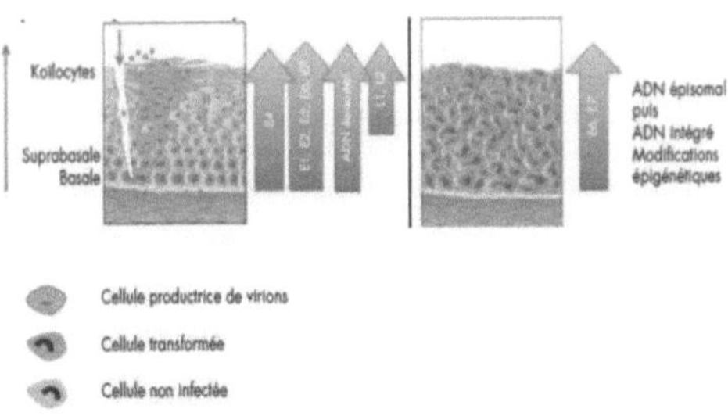

Figure 9: papillomavirus viral cycle (39)

The viral cycle is therefore dependent on cell differentiation. In basal cells, only the early genes are expressed, so no viral particles are present. On the other hand, cells undergoing keratinisation guarantee a complete viral cycle and the production of viral particles which will be released during desquamation (Figure 10).

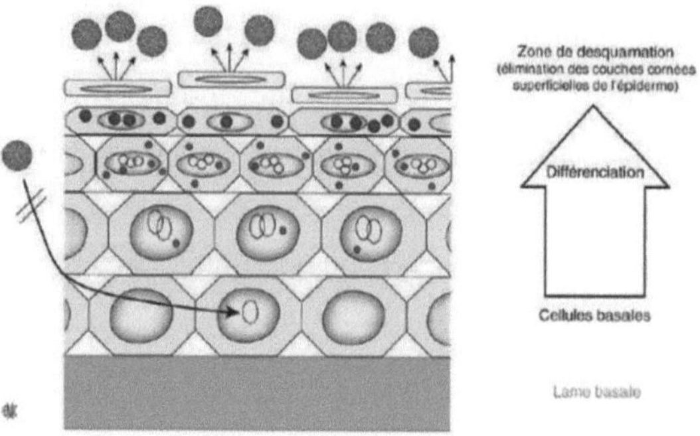

Figure 10: HPV replication parallels cell differentiation (40).

b) Latent infection :

The primary infection goes unnoticed, whatever the type of HPV. Following this primary infection, the virus is carried for around fifteen months. Then, in over 80% of cases, the immune system eliminates the HPV virus. But for the remaining percentage, the infection persists, and the HPV can remain "latent" for several months, leading to lesions and then cancer in the long term.

When the infection remains latent, the viral DNA persists in episomal form (i.e. it is not integrated into the cell's genome). However, under the influence of various endogenous or exogenous factors, the latent state can be broken and the virus can once again develop into a productive infection, leading to relapses.

c) Transforming infection :

In the case of high-risk HPV infection, the viral DNA can integrate the genome of the infected cell. To do this, the viral DNA is cut at the E1/E2 genes and linearised, resulting in a deletion of E1/E2 and a rearrangement of the viral genome. As a result, E2 no longer acts as a repressor of the E6 (linked to the p53 protein) and E7 (linked to the p105RB protein) genes and their transcription is increased. The p53 and p105RB (105kDa retinoblastoma protein) proteins are involved in cell cycle regulation, and are both tumour suppressor proteins. The p53 protein is involved in cell cycle arrest in the G1 phase, leading to cell apoptosis in the presence of non-repairable genomic abnormalities. The p105RB protein is involved in the progression of the cell through the cell cycle depending on its phosphorylation state. It is increasingly phosphorylated as it advances through the cycle during the G1 and S phases, then remains phosphorylated as it passes from the S phase to G2 and is dephosphorylated during the M phase. It then sequesters with transcription factor E2F. Then, during a new cycle, p105RB is phosphorylated again and detached from the E2F factors, which can then exercise their role of activating the genes involved in DNA synthesis. Phosphorylation of the p105RB protein is therefore involved in the cell's passage from the restriction point to the G1 phase. On integration into the cell genome, E6 and E7 are no longer repressed (Figure 11). E6 in the presence of an E6-AP protein (E6 associated protein, present only in HPV HR) will bind and transfer a ubiquitin to the p53 protein, leading to its degradation. As a result, the cell cycle is no longer controlled and genomic abnormalities accumulate, leading to the appearance of a malignant genotype. As for the E7 protein, it has the ability to bind strongly to p105RB, leading to dissociation of the p105RB/E2F complex. E2F is then free and activates the genes involved in DNA synthesis and cell progression. The E7 protein can also bind to cdk/cyclines and p105RB-related proteins.

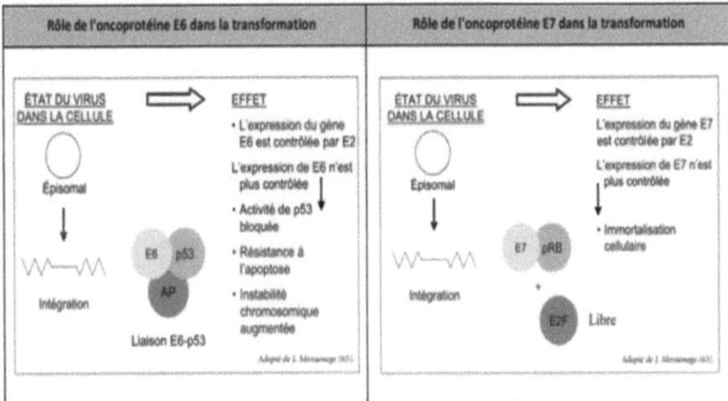

Figure 11: role of onc proteins e6 and e7 in cell transformation (36)

All this then leads to cellular transformation, resulting in the appearance of pre-cancerous or cancerous lesions. Progression to cancer is not systematic, and 32% to 57% of pre-cancerous lesions may regress naturally in a manner opposite to their grade. Cancer develops after several years (around fifteen years) and the transition from pre-cancerous to cancerous lesions is linked to high-grade HPV.

II. LESIONS AND CANCERS ASSOCIATED WITH HPV

HPV-induced lesions are multiple and depend on both the target epithelium and the HPV genotype involved.

⸙ Skin lesions :

HPV-related skin lesions are essentially represented by cutaneous warts and epidermodysplasia verruciformis. Skin warts are benign tumourous lesions that are highly contagious and common (7-10% of the population), occurring mainly in schoolchildren and young adults. They develop mainly on the extremities and face.Transmission can occur through self-inoculation or direct contact with skin. or indirect contact (soiled clothing or soil).The natural course can be favourable, with 60% of cutaneous warts healing spontaneously within 2 years. The HPVs most frequently involved are genotypes 1, 2, 3, 4 and 10. Skin warts are not associated with skin cancer. Epidermodysplasia verruciformis is a rare genetic disease characterised by chronic disseminated HPV infection of the skin. In 80% of cases, the infection is caused by HPV 5 and 8. These HPVs have oncogenic potential, and patients can develop multiple squamous cell carcinomas in photo-exposed areas from the age of 30-40 (41).

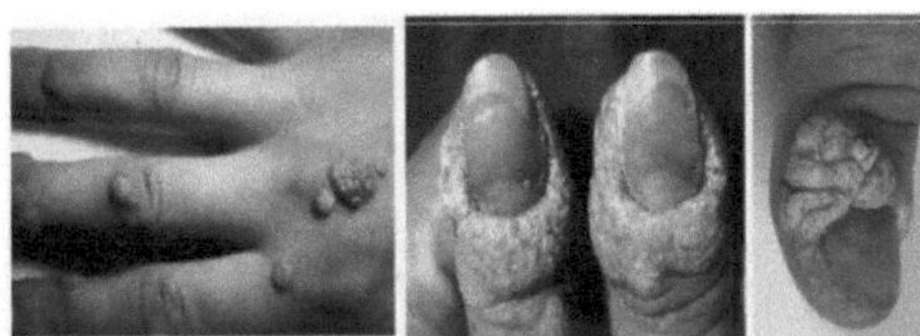

Figure 12: Common warts of the hand caused by HPV (42).

Lesions of the ano-genital mucosa

Condylomata acuminata or genital warts These are benign lesions that appear on the external genitalia (vulva, labia, penis, perineum), vagina, cervix or perianal area.They are caused by infection with mucosal-tropic HPV of low oncogenic risk, HPV 6. and 11 in 90% of cases (43).Transmission occurs through direct contact, mainly sexual (very rarely through neonatal transmission). Prevalence is around 1% of the sexually active population (44). The incubation period for condylomata acuminata is 3 months on average, but may be longer. sometimes reaching several years.Spontaneous evolution or evolution under treatment usually leads to regression. The risk of condyloma acuminata progressing to squamous cell carcinoma is very low (45). Intraepithelial neoplasia Following infection with mucosal-tropic HPVs, mainly of high oncogenic risk, intra-epithelial neoplasia can develop in the cervix (CIN), vagina (VaIN), vulva (VIN), penis or anal canal. Intraepithelial neoplasia or precancerous lesions are lesions characterised by a disorder in the growth and differentiation of the epithelium. By definition, the lesion is strictly intraepithelial and does not cross the basement membrane. These lesions can exist in more or less serious forms and are classified into 3 grades according to the degree of disorganisation of the epithelium: low-grade lesions (CIN 1) and high-grade lesions (CIN 2 and 3).

Table IV: Classification of CIN according to Richer

LÉSION	CIN	HISTOLOGIE
Dysplasie légère (avec koïlocytose)	CIN1	
Dysplasie modérée	CIN2	
Dysplasie sévère Carcinome *in situ*	CIN3	

Cancers caused by high-risk mucosal-tropic HPV A significant proportion of certain cancers are linked to infection by high-risk oncogenic HPV (46,47).

✓ Cervical cancer: 99.7

HPV 16 is by far the most frequent genotype in invasive cervical cancers, with a prevalence of 73%, followed by HPV 18 (19%), 31 (7%) and 33 (4%).

Genotypes 16 and 18 cause around 70% of cervical cancers

✓ Anal cancer: 90% of cases

Over 90% of anal cancers are attributable to HPV 16 and 18.

- Vulvar and vaginal cancer: 40 to 60%.

80% of vulvar and vaginal cancers are attributable to HPV 16 and 18.

✓ Penile cancer: 40% of cases

60% of penile cancers are attributable to HPV 16 and 18.

HPVs have also been found in extra-genital mucous membranes such as the oral cavity, oropharynx and larynx, where they are also implicated in 10 to 30% of cancers. Over 90% of these cancers are linked to HPV 16 and 18.

III. ENITAL INFECTION WITH THE HUMAN PAPILLOMA VIRUS

• **Transmission :**

Genital HPV infection is the most common sexually transmitted infection (STI) in the world (48).

Any sexual act, with or without penetration, is associated with a risk of HPV infection (48), so adolescents who engage in sexual practices such as touching or foreplay, even without penetrative sex, are at risk (49).

A 2006 Canadian study estimated the probability of HPV transmission during sexual intercourse at an average of 40% if one of the partners is a carrier.

(50). This figure, much higher than for other viral STIs (HIV, HSV-2 and HBV infections), is close to the probability of transmission for bacterial STIs (Chlamydia trachomatis, Neisseria gonorrhoeae and Treponema pallidum infections).

HPVs are therefore highly transmissible viruses. It is estimated that 70% of sexually active men and women will contract an HPV infection at least once in their lifetime (51).

Although it helps to reduce the risk of transmission, the protective power of the condom remains controversial, as the virus can be present throughout the anogenital region (including areas not covered by the condom). Nevertheless, according to a 2006 American study, systematic use of condoms during penetrative sex reduces the incidence of infection by 60% (52).

Other modes of transmission have also been described. Vertical maternal-foetal transmission during vaginal delivery is possible. The most common example is juvenile laryngeal papillomatosis, which affects children under the age of 5. This involves transmission of HPV 6 or 11 during vaginal delivery by a mother who is a carrier of genital condylomata acuminata, at the time of passage through the genital tract [53,54]. Finally, horizontal transmission is much rarer, but also possible. HPV viruses are extremely resistant to environmental conditions. This possibility of transmission underlines the importance of strict hygiene rules, particularly for the use of medical equipment for cervical pathology, and the use of single-use gloves and equipment(55). All in all, it is important to specify that HPV cannot be transmitted via blood, semen, saliva or breast milk.

• Prevalence

According to the World Health Organisation, HPV infections affect 660 million people worldwide and the number of new genital infections in women is estimated at 30 million per year (56).

⬇ Disparities in prevalence between countries

The epidemiological study conducted by the International Agency for Research on Cancer (IARC) on more than 1,500 women from 11 countries shows that, all ages combined, the worldwide prevalence of genital HPV infections (all types combined) is 10.5%. The study also found wide variations in prevalence between the countries studied, ranging from 1.4% in Spain to 25.6% in Niger (57). Prevalence, all ages and all HPV types combined, is estimated at 25.6% in Africa, 8.7% in Asia, 14.3% in South America and 5.2% in Europe.

⬇ Disparity in prevalence according to HPV type

Among HPV infections, it would appear that the prevalence is higher for high-risk oncogenic HPVs than for low-risk HPVs. According to the same IARC study, the worldwide prevalence of high-risk HPV infections is 7.6%, compared with 2.9% for low-risk HPV. Of the HPV-positive women identified, 66.8% were infected with high-risk HPV and 27.7% with low-risk HPV. Of the women infected with high-risk HPV, 19.7%, 7.6%, 7.5% and 7.2 respectively were infected with genotypes 16, 58, 31 and 18.

IV. RISK FACTORS

Numerous studies have identified the main factors favouring the transmission of HPV infection. Three main risk factors for HPV infection have been described (58):

❖ Early sexual activity (risk multiplied by 2 if sexual activity before the age of 17). The extension of the endocervix-exocervix junction zone in adolescent girls could explain the susceptibility of this age group to infection when they first have sex.

❖ Multiple sexual partners (increase in risk proportional to the number of sexual partners) partners).

❖ Smoking (risk multiplied by 3): excretion of hydrocarbons in the cervical mucus and local toxicity.

❖ Estrogen-progestogen oral contraception.

❖ The presence of another STI and smoking are also considered risk factors, as they reduce the immune protection against penetration. of HPV and/or by weakening the epithelia

❖ Immunosuppression: HIV infection; long-term corticosteroid therapy; treatment by immunosuppressants.

❖ Low socio-economic status.

V. HPV VACCINATION

1. Vaccine mechanisms of action

HPV vaccination is referred to as a prophylactic vaccine, meaning that it is indicated for the primary prevention of HPV infections. Prophylactic HPV vaccines are based on the discovery in the 90s of the viral pseudo-particle self-assembly properties of the major capsid protein o f HPV viruses (L1 protein). The L1 protein has the spontaneous ability to self-aggregate to form an envelope. a spherical virus-like particle known as a Virus-Like Particle (VLP). VLPs have a morphology almost identical to that of the virus and trick the immune system into producing a high level of specific neutralising antibodies. VLPs are produced by genetic engineering in vitro, by introducing the L1 gene into various eukaryotic cells (insect cells, yeast). The serum antibodies produced in response to the injection of VLPs are directed against the L1 protein of the virus. They are then transuded from the serum to the cervical mucosa where they can neutralise the virions before they enter the basal cells of the epithelium by binding to the L1 protein of the viral capsid. This process prevents the acquisition of HPV infection. These pseudoparticles have neither infectious nor oncogenic potential.

2. Vaccines available

Two HPV vaccines have been developed based on the VLP manufacturing principle: Gardasil® and Cervarix® (Figure 13) (59).

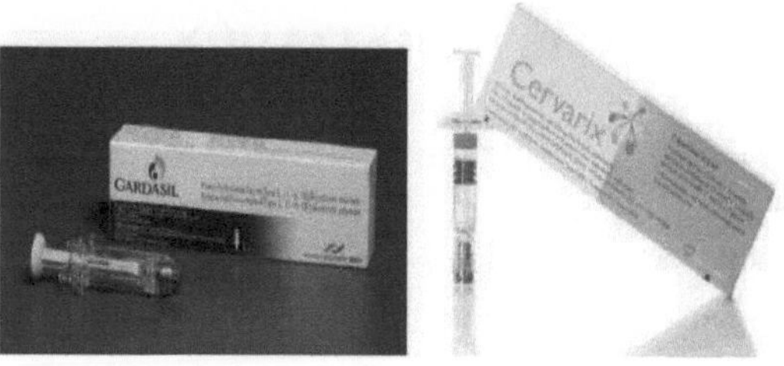

Figure 13: The two HPV vaccines available on the market

Gardasil® is a quadrivalent vaccine developed by Sanofi Pasteur MSD, which received its European Marketing Authorisation in September 2006 and has been marketed in France since November 2006. Cervarix® is a bivalent vaccine developed by the GlaxoSmithKline laboratory. It obtained its European marketing authorisation in September 2007 and has been marketed in France since March 2008. Both vaccines are administered intramuscularly. The two vaccines differ in their composition in terms of active substance and adjuvant (table 5).

Table V: Main characteristics of the 2 HPV vaccines (59)

Trade name (Laboratory)	Gardasil® (merck)	Cervarix® (GSK)
VLP1 content per dose	Type 6: 20 µg Type 11: 40 µg Type 16: 40 µg Type 18: 20 µg	Type 16: 20 µg Type 18: 20 µg
Additive	Hydroxyphosphate amorphous aluminium	AS04 combining an aluminium hydroxide salt

3. Indications and recommendations

Gardasil® is indicated from the age of 9 for the prevention of high-grade genital precancerous lesions of the cervix (CIN 2, CIN 3), vulva (VIN 2, VIN 3) and vagina (VaIN 2, VaIN 3), for the prevention of cervical cancer and genital warts (condyloma acuminata) due to HPV 6, 11, 16 and 18 (60). Cervarix® is indicated from the age of 9 for the prevention of high-grade genital precancerous lesions of the cervix (CIN 2, CIN 3) and cervical cancer due to HPV 16 and 18 (61).In accordance with the opinion of the Haut Conseil de la Santé Publique (HCSP) of 28 September 2012, vaccination against HPV infections is recommended for young girls aged between 11 and 14, with catch-up vaccination limited to the age of 19 (62).

<h1 style="text-align:center">VI. THE P53 PROTEIN</h1>

The transcription factor P53 was first described in 1979. Since then, this protein has been the subject of numerous studies because of its involvement in various biological processes such as :

⬥ Cell blocks in the cell cycle,

⬥ DNA damage repair,

⬥ The development of apoptosis,

⬥ Cellular differentiation.

I.   Structure of P53

The p53 protein encodes a TP53 gene located on the short arm of chromosome 17 (17p13). It is a nuclear phosphoprotein of 53kDa and 393 amino acids divided into 5 domains (63) (figure 1). The main domains are :

❖ The N terminal, a transactivator domain

❖ The DNA-binding domain in the centre

❖ The C-terminal (COOH), a region regulating protein activity

The N-terminal domain contains 2 TAD (Transcription Activation Domain) domains, TAD I and TAD II, as well as a proline-rich domain: PRD (Prolin Rich Domain) (64;65). The TAD I and TAD II domains allow the recruitment of proteins involved in the transcription initiation complex, such as the TFIID factor, or proteins that co-activate transcription, such as the CBP/p300 protein, or proteins involved in its negative feedback control, such as MDM2 (66). The proline-rich domain is composed of several repeats of the PXXP motif. It plays a role in regulating the transcriptional and non-transcriptional activities of p53. Finally, the N-terminal domain contains a nuclear export sequence (NES).

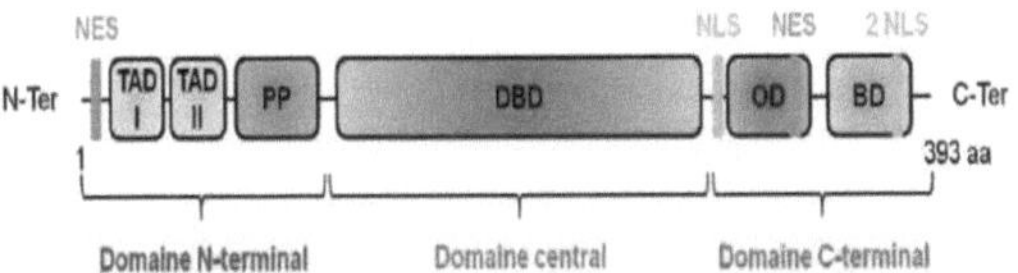

Figure 14: Structure of p53.

The p53 protein is composed of an N-terminal domain comprising the TADI and TADII as well as the proline-rich domain PP, a central domain comprising the DNA binding domain (DBD) and finally a C-terminal domain comprising the oligomerisation domain (OD) and the regulatory domain rich in basic amino acids (BD). The p53 protein has 2 NES and 3 NLS which regulate its nucleocytoplasmic localisation (67).

The central domain of p53 contains the DNA binding domain (DBD). This domain is essential for the transcriptional activities of p53. Certain amino acids in this domain can undergo post-translational modifications that regulate the activities of p53. Furthermore, it has been shown that more than 80% of p53 mutations found in cancers affect the DBD domain (68).

The C-terminal domain contains the OD (Oligomerisation Domain). This domain enables the formation of a p53 tetramer which then binds to the promoters of its target genes (69). Finally, the C-terminal domain also includes a regulatory domain rich in basic amino acids: the BD (Basic Domain). The N-terminal domain also contains 3 NLS sequences and an NES sequence.

II. Cellular functions of P53 :

As a tumour suppressor, the p53 protein plays an essential role in maintaining genome integrity (hence its nickname "genome guardian") in order to prevent inappropriate cell proliferation. Its activity is Involved in the regulation of processes such as the cell cycle, apoptosis and DNA repair (70;71). P53 is also involved in metabolic, autophagic and angiogenic processes.

1. Cell cycle control

The cell cycle is made up of 4 distinct phases: the G1 latency phase, the S DNA replication phase, the G2 latency phase and the M mitosis phase. Control points provided by Cyclin/Cdk (Cyclin-dependent kinase) complexes regulate the transition from one phase to another.Depending on cellular damage, p53 can induce a transient arrest of the cell cycle in G1 phase, preventing damaged DNA from being replicated, an arrest in G2 phase allowing suspension of the cycle before mitosis or an irreversible arrest of the cell cycle leading to senescence.

a) Phase stop G1

This function of p53 involves the activation of the expression of its target gene CDKN1A (Cyclin-Dependent Kinase Inhibitor 1A) encoding the p21 protein, the principal regulator of the G1/S passage point (72). In fact, p21 belongs to the family of CDK inhibitors and blocks the cell cycle in G1 phase by binding to and therefore inactivating the complex (cyclinE/CDK2), thus preventing phosphorylation of the Rb protein and release of E2F, which is necessary for entry into S phase (Figure 15).

b) Phase stop G2

Cell cycle arrest in the G2 phase is the result of destabilisation of the complex (cyclinB1/CDC2), a key regulator of the G2/M crossing point. Indeed, p53 induces the expression of GADD45 (Growth Arrest and DNA-Damage-inducible protein) which binds to CDC2 (Cell Division Cycle protein 2, also known as CDK1) preventing the formation of the (cyclinB1/CDC2) complex (73). In addition, p53 induces the expression of 14-3-38 (74), which prevents the activation of CDC2 by sequestering its activating kinase CDC25.

c) Senescence

The senescence effector programme comprises the p53/p21 and Rb/p16 pathways which irreversibly block cell cycle checkpoints.

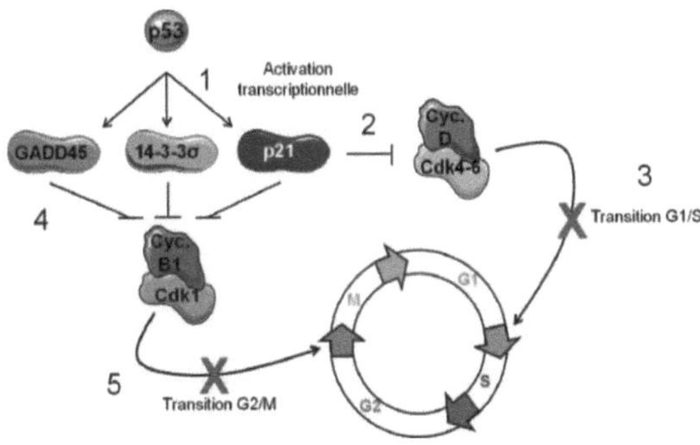

Figure 15: p53 induces cell cycle arrest.

p53 activates transcription of genes encoding GADD45, 14-3-3a and p21 proteins [1]. They are inhibitors of Cyclin-Cdk complexes which regulate transitions between the different phases of the cell cycle. p21 inhibits the Cyclin D/Cdk4-6 complex [2] which controls the G1/S transition [3]. p21, like the GADD45 and 14-3-3a proteins, also inhibits the Cyclin B1/Cdk1 complex [4] which regulates the G2/M transition [5]. (73 ;74)

2. Repair of DNA

There are numerous DNA repair mechanisms, such as MMR (Mismatch Repair), NHEJ (Non Homologous End-Joining), NER (Nucleotide Excision Repair) and BER (Base Excision Repair) (75).NER: p53 can induce the expression of XPC (Xeroderma Pigmentosum group C) and DDB2 (DNA Damage-Binding protein 2) involved in the detection of DNA damage and bind and therefore modulate the activity of XPB and XPD (Xeroderma Pigmentosum B and D) (ATPase and helicase respectively), two sub-units of TFIIH (Transcription Factor II Human). excision effector (76). P53 also induces expression of GADD45 and PCNA (Proliferating Cell Nuclear Antigen), which are involved in NER (77). BER: p53 can interact with components of the BER system such as APE1

(Apurinic/apyrimidinic endonuclease 1) and DNA polymerase B (78). HR: p53 can interact with RAD51 and RAD54, two major components of homologous recombination (79).

3. Apoptosis

Apoptosis is cell death programmed to eliminate cells that are too damaged ("death threshold" model) and is essential for the development and survival of multicellular organisms. There are two main pathways for inducing apoptosis: the intrinsic or mitochondrial pathway, and the extrinsic or death domain receptor pathway. During the extrinsic pathway, activated for example during the immune response, p53 can transactivate the expression of death receptors located on the plasma membrane such as DR4, DR5 and CD95/Fas, which initiate apoptosis by activating caspase 8 on binding of their ligand (TRAIL and CD95L/FasL). Expression of the transmembrane protein PERP (p53 apoptosis effector related to PMP-22) can also be transactivated in response to extensive DNA damage (80).P53 can activate the intrinsic pathway by inducing the expression of mitochondrial pro-apoptotic proteins: BAX (BCL-2-associated X), NOXA (or phorbol-12-myristate- 13-acetate-induced protein 1), PUMA (p53 upregulated modulator of apoptosis), BID (BH3 interacting-domain death agonist), BAD, p53AIP1 (p53-regulated apoptosis-inducing protein 1)... which promote the fall in mitochondrial membrane potential and the release of cytochrome c leading to the formation of an apoptosome with APAF-1 (Apoptotic protease activating factor 1) (whose expression is also induced by p53) and caspase-9. In addition, p53 can repress the expression of the mitochondrial anti-apoptotic proteins BCL-2 and BCL- XL and of apoptosis-inhibiting proteins (IAP family) such as survivin. In addition, p53 can transactivate genes encoding pro-apoptotic proteins located in the cytoplasm, such as: the PTEN protein which opposes the PI3K/AKT survival pathway (81), the PIDD protein (p53-induced protein with a death domain) which is thought to interact with caspase 2 to engage the mitochondrial pathway (82;83), the PIG3 protein (P53-Inducible Gene 3) involved in the induction of apoptosis under conditions of oxidative stress (84) and caspase 6 considered to be an effector caspase (85).

At the same time, and even before this transcriptional response by p53, which takes a certain amount of time to take effect (several hours), p53 can act directly (within a few minutes) in the mitochondria. Indeed, p53 can bind and inhibit the activity of the anti-apoptotic proteins BCL-2 and BCL-XL and therefore facilitate the pro-apoptotic activity of BAX (86). P53 can also bind to BAX, BAD and BAK (BCL-2 homologous antagonist/killer) and directly promote their pro-apoptotic activity (fig 16) (87).

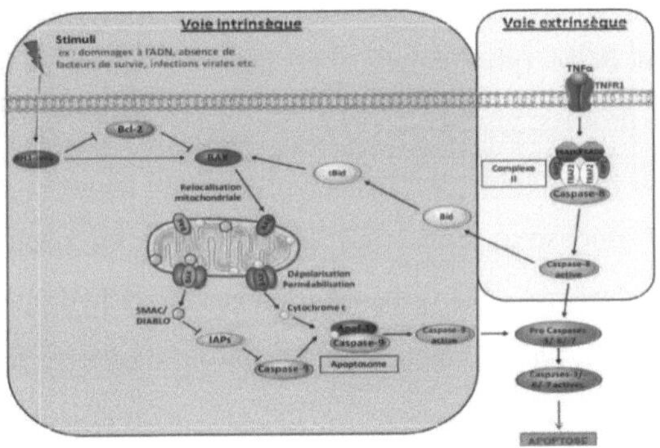

Figure 16: The two main pathways of apoptosis.

There are two main signalling pathways for apoptosis: the extrinsic pathway (yellow), or death domain receptor pathway, and the intrinsic pathway (blue), or apoptosis pathway.mitochondrial pathway. The intrinsic pathway can be activated by various stimuli such as DNA damage, absence of growth factors or viral infection. These two pathways lead to the activation of apoptosis effector proteins called caspases. The extrinsic pathway can join the mitochondrial signalling pathway via the cleavage of Bid into tBid.

4. Metabolism

The p53 protein also plays a role in cell metabolism. In fact, p53 slows down glycolysis by modulating three of its stages:

- Decreased expression of membrane glucose transporters such as GLUT1, GLUT3 and GLUT4, leading to a reduction in the level of intracellular glucose, the initiating source of glycolysis (88; 89).

- Induction of TIGAR (TP53-Induced Glycolysis and Apoptosis Regulator) expression, which inactivates PFK (phosphofructo-kinase), the enzyme involved in the third stage of glycolysis (90).

- Inhibition of PGM (phosphoglycerate mutase), which is involved in the final stages of the process. glycolysis (91).

At the same time, p53 inhibits the pentose phosphate pathway, responsible for the catabolism of glucose to form NADPH and ribose, by inactivating the key enzyme in this pathway: G6PD (glucose-6-phosphate dehydrogenase) (Jiang et al., 2011). In addition, p53 stimulates mitochondrial oxidative phosphorylation. Indeed, p53 increases the expression of SCO2 (synthesis of cytochrome c oxidase 2), thus promoting the activity of the COX complex (cytochrome c oxidase), the fourth complex of the mitochondrial respiratory chain (92). Other p53 targets are indirectly involved in the activation of mitochondrial oxidative phosphorylation: GLS2 (glutaminase 2) which, among other things, promotes the production of glutamate and a-ketoglutarate (93), the Parkine protein which activates the PDH (pyruvate dehydrogenase) complex required for the conversion of pyruvate into acetyl-CoA (94), transrepression of PDK2 (pyruvate dehydrogenase kinase 2) which inhibits the activity of the PDH complex (95), and increased expression of p53R2 ribonucleotide reductase. which helps maintain mitochondrial DNA integrity (96). The activation of mitochondrial oxidative phosphorylation by p53 could be an increase in the production of endogenous ROS, but the system is finely regulated as p53 also activates antioxidant defence processes to counteract any deleterious effects of these ROS. Under physiological conditions and at low levels of activation, p53 exerts antioxidant activity involving the transactivation of proteins such as GPX1 (glutathione peroxidase 1), SOD2, sestrins 1 and 2, GLS2, TP53INP1 (Tumor protein p53-Inducible Nuclear Protein 1) and ALDH4 (aldehyde dehydrogenase 4). On the other hand, under conditions of high intensity stress, p53 appears to exert a pro-oxidant activity by inducing the expression of proteins such as PIG3, proline oxidase and FDRX (ferredoxin reductase), thus participating in cell death by induced apoptosis when cell damage is too great (97) (Figure 17).

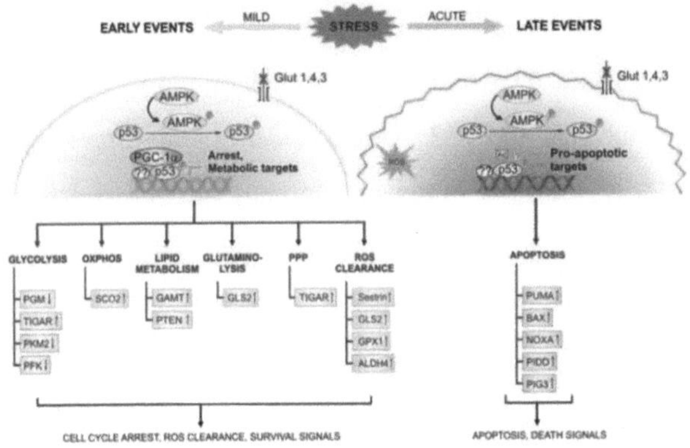

Figure 17: Schematic illustration of the variability of the cellular functions of p53, depending on the intensity of stress.

This diagram illustrates the consequences of a metabolic stress which activates p53 in an AMPK-dependent manner. The left-hand side of the diagram shows an example of a low intensity, short duration stress that activates the survival functions of p53; the right-hand side of the diagram shows a high intensity and/or short duration stress that activates the survival functions of p53. which leads to the induction of p53-dependent apoptosis (98).

5. Autophagy

The term autophagy comes from the Greek words Auto (alone) and Phagy (to eat) and defines a cell that eats itself. It can be beneficial and protective when it enables damaged macromolecules or mitochondria to be eliminated. Under conditions of stress, the autophagic process can be intensified to enable the cell to adapt by, for example, participating in the production of energy from the recycling of products degraded during nutrient deprivation. On the other hand, this intensification of autophagy may also represent a survival mechanism for tumours exposed to a hypoxic microenvironment or to chemotherapy (99). In this sense, p53 plays a dual role depending on its intracellular location (100). Present in the nucleus, the p53 protein induces the expression of proteins involved more or less directly in the activation of autophagy: DRAM (Damage-Regulated Autophagy Modulator) and DAPK1 (Death-Associated Protein Kinase 1), which may also be involved in apoptosis; and AMPK (AMP-activated protein Kinase), TSC1/2 (Tuberous Sclerosis), sestrins and PTEN, which lead to the inactivation of mTOR, thereby lifting the inhibition that mTOR exerts on the autophagic process (100). On the other hand, the p53 protein, which is present in the cytoplasm under basal conditions, appears to exert an inhibitory activity on autophagy by a mechanism that remains to be elucidated (101).

6. Angiogenesis

The p53 protein has anti-angiogenic activities that prevent tumour extension by at least three mechanisms (102):

► Transactivation of anti-angiogenic factors such as TSP-1 (thrombospondin-1), EPHA2 (ephrin receptor A2) and BAI1 (brain-specific angiogenesis inhibitor 1).

► Transrepression of pro-angiogenic factors such as VEGF (103), FGF2 and

FGF2-BP.

► Inhibition of hypoxia detection systems by inducing degradation of HIF-1a (104).

III. The causes and consequences of altered P53 expression :

Alterations in the p53 gene are found in more than half of human cancers, and its central domain (102-292) alone accounts for 90% of mutations. Inactivation of p53 is often caused by missense point mutations in certain key codons, particularly codons 175, 248 and 273.These are considered to be the most frequently mutated codons in the p53 gene sequence and are implicated in the emergence of many cancers.(105). Among the various mechanisms of inactivation of the p53 gene and its protein which have been described, precisely in the sense of a loss of their function, we will mention six which seem to us to be the most important:☐inactivation of the p53 gene by deletion of one or both of its alleles, which reduces or inhibits tetramer formation and leads to a reduction in the expression of target genes.☐inactivation of the p53 protein by :
- nonsense mutations and mutations within the splice site producing proteins truncated and unable to oligomerise
- missense mutations of central domain amino acids. These substituted amino acids can be divided into two groups: a first group comprising the residues involved in maintaining the three-dimensional structure of the DNA-binding domain and a second group comprising residues in direct contact with DNA (106).

These two categories of mutations contribute directly to the generation of proteins for which the ability to bind to DNA is partially or even totally abolished. its binding to certain viral proteins such as the E6 protein of human papillomavirus (HPV16 and 18) and the T antigen of simian virus 40 (SV40)

(107).

- its signalling pathway, which can be disrupted by altering gene expression

MDM2 (double-minute) which is overexpressed and/or amplified (108).

- failure of the translocation mechanisms leading concomitantly to abnormal cytoplasmic sequestration and nuclear exclusion (109).

Particular forms of p53 mutations can give rise to 'gain o f function' mutants. The presence of these mutants in cells gives rise to a new phenotype. The cells acquire new functions normally absent in cells with the wild-type p53 protein. These "gain of function" mutants have a dominant positive activity and can take part in oncogenic processes (110).

IV. P53 and cancer

The p53 protein (a product of the TP53 tumour suppressor gene), considered to be the guardian of the genome, is activated (stabilised) in response to DNA damage. The p53 transcription factor is the target of several viral oncogenes, such as the SV40 T antigen (111), the E1B protein of adenoviruses, EBNA-5 of Epstein Barr virus (112), Tax of HTLV-1 (113), and the E6 protein of HPV16 and 18 viruses (114). Although not required for development, the p53 protein is essential for maintaining genome integrity, its inactivation promoting tumour formation (115). Present in low concentrations under normal conditions, it is activated post-translationally under conditions of physiological stress. Its activation induces cell cycle arrest in G1 phase, and even apoptosis in altered cells. In cells transformed by HPV16 or p53 levels are very low. Oncogenic HPVs produce an E6 protein capable of binding to a cellular ubiquitin ligase called E6-associated protein (E6AP) to form the E6-E6AP complex, which results in the ubiquitination of p53, and hence its degradation by the proteasome (Figure 18) (116). This property is specific to so-called high-risk viruses (frequently associated with cancer), and induces deregulation of the cell cycle by these viruses.

The transcriptional activity of p53 is also recorded by a competitive interaction of E6 with the CBP coactivator at its p53 binding site (117).

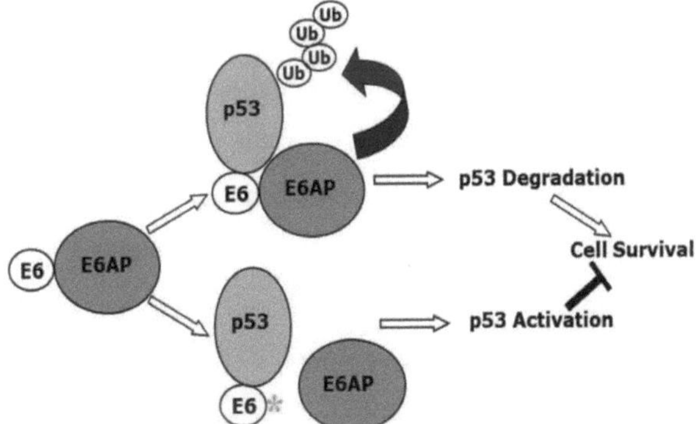

Figure 18: E6, a viral oncoprotein expressed in almost all HPV-positive cancers, mediates the degradation of the tumour suppressor protein p53, a process thatrequires a ternary complex between E6, the E6-associated protein (E6AP) and p53. As a result, the apoptosis-inducing function of p53 is inhibited.

VII. DISCUSSION

According to this bibliographical study, more than 100 types of human HPV have been identified or partially characterised. Although most HPV are responsible for benign skin diseases, numerous studies have demonstrated the major role of certain types of human papillomavirus (HPV) as probable etiological agents in the genesis of cancer and its precursors. Integration of the viral genome leads to over-expression of viral proteins E6 and E7. By interfering with cellular proteins involved in cell cycle control, such a s the P53 protein and the pRb protein, these viral proteins are partly responsible for cell transformation and malignant evolution. The induction of cervical cancer by high-risk HPV, for example, requires the concerted action of two viral oncoproteins, E6 and E7. The E7 oncoprotein is largely responsible for disrupting cell cycle progression by targeting, among others, the pRb family of tumour suppressors (Munger et al., 2001 for review). In turn, the E6 oncoprotein neutralises the cellular response to unscheduled proliferation by targeting p53 for proteasome-mediated degradation (Thomas et al., 1999b). In contrast, a number of studies have shown that the degradation of p53 by E6 requires the recruitment of the cellular ubiquitin ligase E6AP into the complex, allowing the polyubiquitination of p53 and its subsequent degradation by the proteasome (Huibregtse et al., 1993; Scheffner et al., 1993). Although the degradation of p53 is an essential part of the function of E6, it is also clear that other activities of E6 are important in promoting the development of malignant tumours (Nguyen et al., 2003).Furthermore, studies by Liu Y, et all ., (1999) showed that carcinogenesis relies primarily on the expression of two viral oncoproteins, E6 and E7. These act synergistically to immortalise and transform infected cells, partly due to their ability to degrade p53 and Rb respectively.The HR-HPV E6 protein has been shown to lead to ubiquitin-mediated degradation of p53 through direct interaction with the cellular E3 ubiquitin ligase, E6AP (5). The specific action of E6 on p53 is functionally equivalent to the inactivation of p53 by mutation, indicating that the HR-HPV E6/p53 complex represents one of the most important events in cervical carcinogenesis, given the interruption of cell cycle checkpoints and inhibition of apoptosis. The degradation of p53 induced by E6-AP is an effect of of HR-HPV and leads to malignant transformation of epithelial cells. and inactivation of p53. A study published by Hoppe-Seyler et al in (2018) showed that HPV carries two oncogenes, E6 and E7, which are overexpressed in HPV-positive cancers and have transforming and oncogenic properties in tissue culture and animal models. In addition, mutational analyses have shown that the E6 and E7 genes of high-risk HPV are sufficient to immortalise human keratinocytes and fibroblasts (de Sanjose et al., 2007). The HPV16 and HPV18 E6 and E7 oncoproteins target the tumour suppressors p53 and retinoblastoma (pRB), respectively, for ubiquitin-mediated degradation, and induce cell proliferation, cell survival, genome instability and innate immune evasion (Dyson et al., 1989; Scheffner et al., 1990, 1993; Huibregtse et al., 1991, 1993).The E6 oncoproteins of high-risk HPV, but not those of low-risk HPV, interfere with the transcriptional activity of p53 and induce p53 degradation. The E6-associated protein (E6AP), the founding member of the HECT E3 ubiquitin ligase family, has been shown to mediate binding between E6 and p53 (Scheffner et al., 1990, 1993; Huibregtse et al., 1991, 1993). The role of E6AP in E6-mediated p53 degradation has been well characterised in vitro and in vivo: E6, E6AP and p53 form a complex that directs the ligase activity of E6AP towards p53, subsequently triggering p53 degradation via the intracellular ubiquitin-proteasome system (Zanier et al., 2012; Vande Pol and Klingelhutz, 2013; Martinez-Zapien et al., 2016). While the major mechanism of E6-mediated oncogenesis is exerted by its ability to suppress the pro-apoptotic effects of p53, recent studies have demonstrated that E6 binds to several cellular proteins and induces their ubiquitination and degradation in an E6AP-dependent manner (White et al., 2012, Mesri et al., 2014).

CONCLUSION

the human papillomavirus (HPV), whose oncogenic potential lies in the expression o f viral proto-oncogenes E6/E7. The carcinogenic potential of these viral proteins lies essentially in their actions on the products of the p53 and RB tumour suppressor genes. The products of these genes, p53 and RB, are part of the cellular DNA damage response (CDR) signalling pathways and their loss results in a loss of functionality leading to genomic instability. In the long term, and in the presence of other factors, thesewill lead to the development of cancer. Advanced cases of these cancers are mainly treated with radiotherapy and concomitant chemotherapy. Chemo-radiotherapy used as a treatment is effective but results in a high rate of morbidity and a significant number of patients will have a recurrence.

BIBLIOGRAPHY

1. Mansour C. Human papillomaviruses. In: Tyring S, ed. Mucocutaneous manifestations of viral diseases. 2005.

2. de Villiers EM, Fauquet C, Broker TR, Bernard HU, zurHausen H. Classification of papillomaviruses. Virology 2004; 324: 17-27.

3. Ke Zhang1T11T, Zhanjun Liu, Ji Li Analysis of human papilloma virus type 52 integration status in exfoliated cervical cells Experimental and Therapeutic Medicine 2017;14:5817-5824

4. Parkin DM, Bray F. Chapter 2: The burden of HPV-related cancers. Vaccine. 21 August

B2006 ;24, Supplement 3:S11-S25.

5. French National Cancer Institute. Cervical cancer in France: 2010 review, Collection Rapport & synthèse. 2010 Jul.

6. Gavillon N, Vervaet H, Derniaux E, Terrosi P, Graesslin O, Quereux C.

Human papillomavirus (HPV): how did I catch it? Gynaecology

7 G.M. Clifford, R.K. Rana, S. Franceschi, J.S. Smith. Human Papillomavirusgenotype distribution in low-grade cervical lesions: comparison bygeographic region and with cervical cancer. Cancer Epidemiology Biomarkers andPrevention. May 2005;14(5):1157-1164.

8 Lorincz AT, Reid R, Jenson AB, GreenbergMD, Lancaster W, Kurman RJ. Humanpapillomavirus infection of the cervix: relative risk associations of 15 common anogenital types. Obstet Gynecol 1992; 79: 328-37.

9 Linzer D.I.H. and Levine A.J.; Characterization of a 54K Dalton cellular SV40 tumor antigen present in SV40- transformed cells and uninfected embryonal carcinoma cells; 1979 ; Cell ; Vol 17 ; 43-52.

10 Baker SJ, Fearon ER, Nigro JM, Hamilton SR, Preisinger AC, Jessup JM, vanTuinen P, Ledbetter DH, Barker DF, Nakamura Y, White R, and Vogelstein

B; Chromosome 17 deletions and p53 gene mutations in colorectal carcinomas; 1989 ; Science 14;244(4901):217-21.

11 Beutner KR, Ferenczy A. Therapeutic approaches to genital warts. Am J Med 1997; 10: 28-37.

12 HPV, cervical dysplasia and cervical cancer

13 de Villiers EM, Fauquet C, Broker TR, Bernard HU, zurHausen H. Classification of papillomaviruses. Virology 2004; 324: 17-27.

14 Segondy M. Classification des papillomavirus (HPV). Revue francophone des laboratoires. 2008.

15 Bernard HU. The clinical importance of the nomenclature, evolution and taxonomy of humanpapillomaviruses. J Clin Virol 2005; 32 (Suppl 1): S1-6.

16 Berraho M, Obtel M, Bendahhou K, Zidouh A, Errihani H, Benider A, et al. Sociodemographicfactors and delay in the diagnosis of cervical cancer in Morocco. Pan Afr Med J. 2012 ; 12:14. PMID:22826738

17 Bouvard V, Baan R, Straif K, Grosse Y, Secretan B, El Ghissassi F, et al. A review of human carcinogens--Part B: biological agents. The Lancet Oncology. 2009;10(4):321- 2.

18 Komlos KF, Kocjan BJ, Kosorok P, Luzar B, Meglic L, Potocnik M, et al. Tumor- specific and gender-specific pre-vaccination distribution of human papillomavirus types. 6 and 11 in anogenital warts and laryngeal papillomas: a study on 574 tissue specimens. Journal of medical virology. 2012;84(8):1233-41.

19 Seedat RY, Thukane M, Jansen AC, Rossouw I, Goedhals D, Burt FJ. HPV types causing juvenile recurrent laryngeal papillomatosis in South Africa. International journal of pediatric otorhinolaryngology. 2010;74(3):255-9.

20 Muñoz N, Castellsagué X, de González AB, Gissmann L. Chapter 1: HPV in the etiology of human cancer. Vaccine. 2006 Aug 31;24 Suppl 3:S3/1-10

21   Robert François Detection of high-risk HPV as an alternative for women not adherent to cervical cancer cytology: a pilot study on the acceptability and

feasibility of vaginal self-sampling and urine sampling. These phar 2016; University of Poitiers Faculty of Medicine and Pharmacy

22 https://disiciencia.files.wordpress.com/2012/02/papiloma.jpg?w=300

23 https://www.researchgate.net/profile/Thales_Fernandes/publication/22192258 8/figure/ fig1/AS:305268136136704@1449793003194/The-structure-of-HPV-Adapted-from- Swiss-Institute-of-Bioinformatics-Viral-Zone.png

24 **S.** Douvier, S. Dalac, 2004**,** Papillomavirus infections Encyclopédie Médico-Chirurgicale 8-054-A-10; p :2

25 J Bodily, LA Laimins. Trends Microbiol, 2011, 19: 33-39)

26 https://encryptedtbn0.gstatic.com/images?q=tbn:ANd9GcRslPwf0uKraI8nk5 ZPnpF3n m0W7bSJ_EFzb36S1JPeOp07DS8l

27 Tommasino M. The human papillomavirus family and its role in carcinogenesis. Semin Cancer Biol 2014;26:13-21.

28 S Beaudin, M Naspetti, and C Montixi, "Le papillomavirus humain : actualisation des connaissances".

29 C AJ Horvath et al, "Mechanisms of cell entry by human papillomaviruses: an overview", Virology Journal 7 (20 January 2010): 11,

30 C AJ Horvath et al, "Mechanisms of cell entry by human papillomaviruses: an overview", Virology Journal 2010 : 7:11

31 "UCLOUVAIN FDP Virology, "Introduction to virology", accessed 27 December 2018,https://www.virologie-uclouvain.be/fr/chapitres/exemples-selected/papillomavirus

32 F Tessier, "Cours virologie pharmacie - Papillomavirus infections".

33 S Alain, S Hantz, and F Denis, "Papillomavirus: viruses and the pathophysiology of infection", mt paediatrics, vol 13, no 1 (January-February 2012).

34 Direction générale de la santé, Comité technique des vaccinations "Vaccination contre les infections à Papillomavirus humains", Guide des vaccinations, 2012 edition: 149 - 159,

Available at:

https://solidaritessante.gouv.fr/IMG/pdf/Guide_des_vaccinations_editio n_2012.pdf

35 Ch Mougin et al, "Papillomavirus humains, cycle cellulaire et cancer du cerv de l'utérus", Journal de Gynécologie Obstétrique et Biologie de la Reproduction, vol 29, n°1 (February 2000): 13 Available at: http://www.em-consulte.com/en/article/113912.

36 J Monsonego, "Papillomavirus infections: state of knowledge, practices and vaccine prevention" (2007).

37 A C de Freitas, E Campos Coimbra, and M da Conceição Gomes Leitão, "Molecular targets of HPV oncoproteins: Potential biomarkers for cervical carcinogenesis," Biochimica et Biophysica Acta (BBA) - Reviews on Cancer 1845, no. 2 (April 1, 2014): 91-103

38 K Münger et al, "Mechanisms of Human Papillomavirus-Induced Oncogenesis",
Journal of Virology 78, no 21 (November 2004): 11451-11460

39 S Alain, S Hantz, and F Denis, "Papillomavirus: viruses and the pathophysiology of infection", Médecine thérapeutique / Pédiatrie 13, no 1 (1 January 2010): 5-19

40 S Mas, "Oncogenesis and the viral life cycle", available at http://mas.stephanie.free.fr/cours_micro_gf/oncogenese.pdf.

41 Zambruno G. Epidermodysplasia verruciformis [Internet]. Orphanet; 2010. Available from: http://www.orpha.net/consor/cgi-bin/OC_Exp.php?Lng=FR&Expert=302

42 Cubie HA. Diseases associated with human papillomavirus infection. Virology. 2013 Oct;445(1-2):21-34.

43 Mougin C, Nicolier M, Decrion-Barthod AZ. HPV and cancers: mechanisms of oncogenesis. Rev Francoph Lab. 2008

44 Quéreux C, Bory J-P, Graesslin O. Condyloma acuminata [Internet]. Collège

Nationaldes Gynécologues et Obstétriciens Français; 2007. Available from: http://www.cngof.asso.fr/d_livres/2007_GM_027_quereux.pdf

45 Ancelle-Park R, Autran B, Baldauf. Groupe de travail sur la vaccination contre les papillomavirus. Comité Technique Des Vaccinations/Conseil Supérieur d'Hygiène Publique de France; 2007.

46 Parkin DM, Bray F, Ferlay J, Pisani P. Global cancer statistics, 2002. CA Cancer J Clin. 2005 Apr;55(2):74-108.

47 Coursaget P, Touzé A. Vaccines against papillomavirus. Virologie. 2006;10(5).

48 Winer RL, Lee S-K, Hughes JP, Adam DE, Kiviat NB, Koutsky LA. Genitalhuman papillomavirus infection: incidence and riskfactors in a cohort of femaleuniversitystudents. Am J Epidemiol. 2003 Feb 1;157(3):218-26.

49 Moscicki A-B. Impact of HPV infection in adolescent populations. J AdolescHealth Off Publ Soc Adolesc Med. 2005 Dec;37(6 Suppl):S3-9.

50 Burchell AN, Richardson H, Mahmud SM, Trottier H, Tellier PP, Hanley J, et al. Modeling the sexual transmissibilityof human papillomavirusinfectionusingstochasticcomputersimulation and empiricaldatafrom a cohortstudyofyoungwomen in Montreal, Canada. Am J Epidemiol. 2006 Mar 15;163(6):534-43.

51 Mougin C, Dalstein V. Epidemiology, natural history and detection of HPV. Bio Trib Mag. 2004;9(1).

52 Winer RL, Hughes JP, Feng Q, O'Reilly S, Kiviat NB, Holmes KK, et al. Condom use and the risk of genitalhuman papillomavirus infection in youngwomen. N Engl J Med. June 22, 2006;354(25):2645-54.

53 Smith EM, Parker MA, Rubenstein LM, Haugen TH, Hamsikova E, Turek LP, et al. Evidence for Vertical Transmission of HPV fromMothers to Infants, Infect Dis ObstetGynecol. 14 March 2010.

54 Lee SM, Park JS, Norwitz ER, Koo JN, Oh IH, Park JW, et al. Risk of Vertical Transmission of Human Papillomavirus throughoutPregnancy: A

Prospective Study. 8(6).

55 Schiffman M, Castle PE, Jeronimo J, Rodriguez AC, Wacholder S. Human papillomavirus and cervical cancer. Lancet LondEngl. 8 Sep 2007;370(9590):890-907.

56 Monsonego J. Papillomavirus infections. State of knowledge, practices and vaccine prevention. Springer DL. 2006.

57 Clifford GM, Gallus S, Herrero R, Muñoz N, Snijders PJF, Vaccarella S, et al, IARC HPV Prevalence Surveys Study Group. Worldwide distribution of human papillomavirus types in cytologically normal women in the International Agency for Research on Cancer HPV prevalence surveys: a pooled analysis. Lancet Lond Engl. 2005 Sep 17;366(9490):991-8.

58 Burchell AN, Winer RL, de Sanjosé S, Franco EL. Chapter 6: Epidemiology and transmission dynamics of genital HPV infection. Vaccine. 2006 Aug 31;24 Suppl 3:S3/52-61.

59 Vidal dictionary. 91st edition. Vidal; 2015.

60 Commission de la tranparence: opinion of 1 February 2012. Gardasil. Haute Autorité de Santé; 2012.

61 Commission de la tranparence: opinion of 1 February 2012. Cervarix. Haute Autorité de Santé; 2012.

62 Opinion on the revision of the vaccination age against human papillomavirus infections for young girls. Haut Conseil de la santé publique; 2012 Sep.

63 Weston A. and Godbold J. H., Polymorphisms of H-ras-1 and p53 in Breast Cancer and Lung Cancer: A Metaanalysis,1997, Environmental Health Perspectives Vol 105.

64 Fields S, Jang SK. 1990. Presence of a potent transcription activating sequence in the p53 protein. Science 249:1046-1049.

65 Dawson R, Muller L, Dehner A, Klein C, Kessler H, Buchner J. 2003. The N-terminal domain of p53 is natively unfolded. J Mol Biol 332:1131-1141.

66 Wu X, Bayle JH, Olson D, Levine AJ. 1993. The p53-mdm-2 autoregulatory

feedback loop. Genes Dev 7:1126-1132.

67 Millau JF, Bastien N, Drouin R. 2009. P53 transcriptional activities: a general overview and some thoughts. Mutat Res 681:118-133.

68 Hainaut P, Hernandez T, Robinson A, Rodriguez-Tome P, Flores T, Hollstein M, Harris CC, Montesano R. 1998. IARC Database of p53 gene mutations in human tumors and cell lines: updated compilation, revised formats and new visualisation tools. Nucleic Acids Res 26:205-213.

69 Kraiss S, Quaiser A, Oren M, Montenarh M. 1988. Oligomerization of oncoprotein p53.
J Virol 62:4737-4744.

70 Vogelstein B, Lane D, Levine AJ, Surfing the p53 network, Nature 408, 307, 2000

71 Vousden KH, Lu X, Live or let die: the cell's response to p53, Nat Rev Cancer, 2002.

72 el-Deiry WS, Harper JW, O'Connor PM, Velculescu VE, Canman CE, Jackman J, et al (1994). WAF1/CIP1 is induced in p53-mediated G1 arrest and apoptosis. Cancer research 54(5): 1169-1174.

73 Zhan Q, Antinore MJ, Wang XW, Carrier F, Smith ML, Harris CC, et al (1999). Association with Cdc2 and inhibition of Cdc2/Cyclin B1 kinase activity by the p53- regulated protein Gadd45. Oncogene 18(18): 2892-2900.

74 Hermeking H, Lengauer C, Polyak K, He TC, Zhang L, Thiagalingam S, et al (1997). 14-3-3 sigma is a p53-regulated inhibitor of G2/M progression. Molecular cell 1(1): 3- 11.

75 Janicke RU, Sohn D, Schulze-Osthoff K. 2008. The dark side of a tumor suppressor: anti-apoptotic p53. Cell Death Differ 15:959-976.

76 Adimoolam S, Ford JM. 2002. p53 and DNA damage-inducible expression of the xeroderma pigmentosum group C gene. Proc Natl Acad Sci U S A 99:12985-12990.

77 Smith ML, Ford JM, Hollander MC, Bortnick RA, Amundson SA, Seo YR,

et al. (2000). p53-mediated DNA repair responses to UV radiation: studies of mouse cells lacking p53, p21, and/or gadd45 genes. Molecular and cellular biology 20(10): 3705- 3714.

78 Seo YR, Fishel ML, Amundson S, Kelley MR, Smith ML. 2002. Implication of p53 in base excision DNA repair: in vivo evidence. Oncogene 21:731-737.

79 Linke SP, Sengupta S, Khabie N, Jeffries BA, Buchhop S, Miska S, et al. (2003). p53 interacts with hRAD51 and hRAD54, and directly modulates homologous recombination. Cancer research 63(10): 2596-2605.

80 Ihrie RA, Reczek E, Horner JS, Khachatrian L, Sage J, Jacks T, et al (2003). Perp is a mediator of p53-dependent apoptosis in diverse cell types. Current biology : CB 13(22): 1985-1990.

81 Mayo LD, Donner DB (2002). The PTEN, Mdm2, p53 tumor suppressor-oncoprotein network. Trends in biochemical sciences 27(9): 462-467.

82 Bock FJ, Peintner L, Tanzer M, Manzl C, Villunger A (2012). P53-induced protein with a death domain (PIDD): master of puppets? Oncogene 31(45): 4733-4739.

83 Oliver TG, Meylan E, Chang GP, Xue W, Burke JR, Humpton TJ, et al (2011). Caspase-2-mediated cleavage of Mdm2 creates a p53-induced positive feedback loop. Molecular cell 43(1): 57-71.

84 Kotsinas A, Aggarwal V, Tan EJ, Levy B, Gorgoulis VG (2012). PIG3: a novel link between oxidative stress and DNA damage response in cancer. Cancer letters 327(1-2): 97-102.

85 MacLachlan TK, El-Deiry WS (2002). Apoptotic threshold is lowered by p53 transactivation of caspase-6. Proceedings of the National Academy of Sciences of the United States of America 99(14): 9492-9497.

86 Mihara M, Erster S, Zaika A, Petrenko O, Chittenden T, Pancoska P, et al. (2003). p53 has a direct apoptogenic role at the mitochondria. Molecular cell 11(3): 577-590.

87 Chipuk JE, Kuwana T, Bouchier-Hayes L, Droin NM, Newmeyer DD,

Schuler M, et al (2004). Direct activation of Bax by p53 mediates mitochondrial membrane permeabilization and apoptosis. Science 303(5660): 1010-1014.

88 Kawauchi K, Araki K, Tobiume K, Tanaka N (2008). p53 regulates glucose metabolism through an IKK-NF-kappaB pathway and inhibits cell transformation. Nature cell biology 10(5): 611-618.

89 Schwartzenberg-Bar-Yoseph F, Armoni M, Karnieli E (2004). The tumor suppressor p53 down-regulates glucose transporters GLUT1 and GLUT4 gene expression. Cancer research 64(7): 2627-2633.

90 Bensaad K, Tsuruta A, Selak MA, Vidal MN, Nakano K, Bartrons R, et al (2006).TIGAR, a p53-inducible regulator of glycolysis and apoptosis. Cell 126(1): 107-120.

91 Corcoran CA, Huang Y, Sheikh MS (2006). The regulation of energy generating metabolic pathways by p53. Cancer biology & therapy 5(12): 1610-1613.

92 Matoba S, Kang JG, Patino WD, Wragg A, Boehm M, Gavrilova O, et al. (2006). p53 regulates mitochondrial respiration. Science 312(5780): 1650-1653.

93 Hu W, Zhang C, Wu R, Sun Y, Levine A, Feng Z (2010). Glutaminase 2, a novel p53 target gene regulating energy metabolism and antioxidant function. Proceedings of the National Academy of Sciences of the United States of America 107(16): 7455-7460.

94 Zhang C, Lin M, Wu R, Wang X, Yang B, Levine AJ, et al. (2011a). Parkin, a p53 target gene, mediates the role of p53 in glucose metabolism and the Warburg effect. Proceedings of the National Academy of Sciences of the United States of America 108(39): 16259-16264.

95 Contractor T, Harris CR (2012). p53 negatively regulates transcription of the pyruvate dehydrogenase kinase Pdk2. Cancer research 72(2): 560-567.

96 Bourdon A, Minai L, Serre V, Jais JP, Sarzi E, Aubert S, et al (2007). Mutation of RRM2B, encoding p53-controlled ribonucleotide reductase (p53R2), causes severe mitochondrial DNA depletion. Nature genetics 39(6):

776-780

97 Vousden KH, Prives C (2009a). Blinded by the Light: The Growing Complexity of p53.Cell 137(3): 413-431.

98 98 Sen N, Satija YK, Das S (2012). p53 and metabolism: old player in a new game.Transcription 3(3): 119-123.

99 Morselli E, Galluzzi L, Kepp O, Vicencio JM, Criollo A, Maiuri MC, et al (2009). Anti- and pro-tumor functions of autophagy. Biochimica et biophysica acta 1793(9): 1524-1532.

100 Maiuri MC, Galluzzi L, Morselli E, Kepp O, Malik SA, Kroemer G (2010). Autophagy regulation by p53. Current opinion in cell biology 22(2): 181-185.

101 Vousden KH, Ryan KM (2009b). p53 and metabolism. Nature reviews. Cancer 9(10): 691-700.

102 Teodoro JG, Evans SK, Green MR (2007). Inhibition of tumor angiogenesis by p53: a new role for the guardian of the genome. J Mol Med (Berl) 85(11): 1175-1186.

103 Mukhopadhyay D, Tsiokas L, Sukhatme VP (1995). Wild-type p53 and v-Src exert opposing influences on human vascular endothelial growth factor gene expression. Cancer research 55(24): 6161-6165.

104 Ravi R, Mookerjee B, Bhujwalla ZM, Sutter CH, Artemov D, Zeng Q, et al (2000). Regulation of tumor angiogenesis by p53-induced degradation of hypoxia- inducible factor 1alpha. Genes & development 14(1): 34-44.

105 Hainaut P, Hollstein M. P53 and human cancer: the first ten thousand mutations. Adv Cancer Res 2000; 77: 81-137.

106 Cho Y, Gorina S, Jeffrey PD, Pavletich NP. Crystal structure of a p53 tumor suppressor-DNA complex: understanding tumorigenic mutations. Science 1994 ; 265 : 346-55.

107 Munger K, Scheffner M, Huibregtse JM, Howley PM. Interactions of HPV E6 and E7 oncoproteins with tumour suppressor gene products. Cancer Surv 1992; 12: 197-217.

108 Oliner JD, Pietenpol JA, Thiagalingam S, et al. Oncoprotein MDM2 conceals the activation domain of tumour suppressor 53. Nature 1993 ; 29 : 857-60.

109 Moll UM, Ostermeyer AG, Haladay R, et al. Cytoplasmic sequestration of wild- type p53 protein impairs the G1 checkpoint after DNA damage. Mol Cell Biol 1996; 3: 1126-37.

110 Dittmer D, Pati S, Zambetti G, et al. Gain of function mutations in p53. Nat Genet1993; 4: 42-6.

111 Lane, D.P. and Crawford, L.V. (1979) T antigen is bound to a host protein in SV40-transformed cells. Nature, 278, 261-263.

112 Szekely, L., Selivanova, G., Magnusson, K.P., Klein, G. and Wiman, K.G. (1993) EBNA-5, an Epstein-Barr virus-encoded nuclear antigen, binds to the retinoblastoma and p53 proteins. Proc Natl Acad Sci U S A, 90, 5455-5459.

113 Uittenbogaard, M.N., Giebler, H.A., Reisman, D. and Nyborg, J.K. (1995) Transcriptional repression of p53 by human T-cell leukemia virus type I Tax protein. J Biol Chem, 270, 28503-28506.

114 Werness, B.A., Levine, A.J. and Howley, P.M. (1990) Association of human papillomavirus types 16 and 18 E6 proteins with p53. Science, 248, 76-79.

115 Donehower, L.A. and Bradley, A. (1993) The tumor suppressor p53. Biochim Biophys Acta, 1155, 181-205.

116 Huibregtse, J.M., Scheffner, M. and Howley, P.M. (1993) Cloning and expression of the cDNA for E6-AP, a protein that mediates the interaction of the human papillomavirus E6 oncoprotein with p53. Mol. Cell. Biol. 13, 775-784.

117 Patel, D., Huang, S.M., Baglia, L.A. and McCance, D.J. (1999) The E6 protein of human papillomavirus type 16 binds to and inhibits co-activation by CBP and p300. EMBO J., 18, 5061-5072.

Printed by Books on Demand GmbH, Norderstedt / Germany